Table 1.1 Threat matrix as a function of modes (modified from Ramirez-Marquez, 2007)

Mode	Consequence			Target				
	Threaten	Disrupt	Destroy/ Damage	Economy	People	Symbolic	Critical Infra-structure	Environ-ment
Physical - Intended								
Spies		X	X			X	X	
Terrorists	X	X	X		X	X	X	
Criminals	X	X		X				
Vandals	X	X	X			X	X	X
Enemies		X	X	X	X	X	X	
Disgruntled	X	X	X			X	X	X
Physical - Unintended								
Software		X					X	
Human		X					X	
Organ		X					X	
Info Tech		X					X	
Hardware		X					X	
Cyber - Intended								
Hackers		X				X	X	
Spies		X	X	X		X	X	
Terrorists	X	X	X	X		X	X	X
Criminals		X		X		X		
Vandals	X	X	X			X	X	
Disgruntled	X	X		X		X		
Enemies	X	X	X	X		X	X	X
Cyber - Unintended								
Software		X		X		X	X	
Human		X		X		X	X	
Organ		X		X		X	X	
Info Tech		X		X		X	X	
Hardware		X		X		X	X	
Natural								
Hurricanes	X	X	X	X			X	X
Earthquake		X	X	X			X	X
Floods	X	X	X	X			X	X
Tornados		X	X	X			X	X
Drought	X	X	X				X	X
Fire	X	X	X				X	X
Volcanoes		X	X				X	X

In today's resource-constrained environment, the Army must exercise wise stewardship of every dollar it manages. A key element in our stewardship is to develop and use sound CBA practices throughout all requirement/resourcing processes. For every proposed program, initiative or decision point that will be presented to decision-makers, it is important to provide an accurate and complete picture of both the costs estimates and the benefits to be derived.

A CBA provides decision-makers with facts, data, and analysis required to make an informed decision. Specifically, a CBA:
- Is a decision support tool that documents the predicted effect of actions under consideration to solve a problem or take advantage of an opportunity.
- Is a structured proposal that functions as a decision package for organizational decision-makers.
- Defines a solution aimed at achieving specific Army and organizational objectives by quantifying the potential financial impacts and other business benefits such as:
  o Savings and/or cost avoidance,
  o Revenue enhancements and/or cash-flow improvements,
  o Performance improvements, and
  o Reduction or elimination of a capability gap
- Considers all benefits to include non-financial or difficult to quantify benefits of a specific course of action (COA) or alternative.
- An analysis of needs and problems, their proposed alternative solutions, and a risk analysis to lead the analyst to a recommended choice before a significant amount of funds are invested by the stakeholders.
- Must be tailored to fit the problem, because finding the optimal solution is the focus of the CBA.
- Supports the decision making process, but will not make a final decision. That will be the responsibility of the decision maker/leadership.
- Is not a substitute for sound judgment, management, or control.

The Office of the Secretary of Defense (OSD) as well as the senior leaders of the Department of the Army (DA) have mandated the use of CBAs to support resource decision making. This guide is applicable to a wide range of energy security requirements, issues, tasks, and problems that require a deliberate analysis to arrive at the optimum course of action.

This guide describes a CBA process that comprises eight major steps.
1. Define the problem / opportunity, to include background and circumstances,
2. Define the scope and formulate facts and assumptions,
3. Define and document alternatives (including the status quo if relevant),
4. Develop cost estimates for each alternative (including status quo if relevant),
5. Identify quantifiable and difficult to quantify benefits,
6. Define alternative selection criteria,
7. Compare alternatives, and
8. Report results and recommendations.

These eight steps are shown graphically in Figure 1.2.

Chapter 1
Purpose, Introduction, and Literature Review

1.1 Purpose

The purpose of this research is to develop and articulate a cost benefit analysis (CBA) methodology specifically for investments for energy security projects for the Department of the Army. This guide will assist Army analysts and agencies in preparing a CBA to support Army decision-makers in funding energy security focused capital projects. This guide will also assist analysts in identifying, quantifying, and evaluating the future costs and benefits of alternative solutions for energy security projects.

Energy security is one component of general physical security of installation that has recently gained great importance because of the fragility of the national power infrastructure. The United States (U.S.) Army's energy security vision is "an effective and innovative Army energy posture, which enhances and ensures mission success and quality of life for our soldiers, civilians and their families through leadership, partnership, and ownership, and also serves as a model for the nation."[1] In the context of this report we are focused solely on energy security for Army installations. Our working definition for energy security is that the "equipment and processes are in place to ensure power is available to an installation to accomplish its critical mission."

This document/guide is intended for mainly in the operational energy security functional areas. Note that this guide was written mainly by modifying the U.S. Army, Cost Benefit Analysis Guide, 3rd Edition dated 22 February 2012. That document was prepared by the Office of the Deputy Assistant Secretary of the Army (Cost and Economics) (U.S. Army, 2012). We simply edited that document to make it energy security specific.

All military operations require energy, and how the armed forces use this "operational energy" can enhance or undermine military effectiveness. Nonetheless, it is new for Department of Defense (DoD) components to consider operational energy as a distinct program or capability, rather than a commodity that can be included in military planning as an assumption. As described in the 2010 Quadrennial Defense Review (QDR), DoD energy security means having assured access to reliable supplies of energy and the ability to protect and deliver sufficient energy to meet operational needs. It is implicit in this definition that military energy security enhances and does not sacrifice other operational capabilities.[2] We are limiting our research and this publication to the operational needs of military installations.

1.2 Introduction

If the U.S. were attacked from an external, sophisticated enemy or even an internal, disgruntled employee, the energy grid would be an easy target to damage/destroy that would lead to catastrophic results. This could cut off power to military bases and greatly degrade our ability to project forces in addition to crippling the U.S. economy. Figure 1.1 graphically shows how power is currently distributed to most military installations. Creative economics, to include 3rd party financing, and engineering solutions are needed to finance these for strategic instruments to ensure the security of our installations, but to also comply with the myriad of regulations, orders, and laws that govern how they are operated. Also, as the largest single user of power and fuel, DoD must take a leadership role in shaping the national energy debate and policy.

---

[1] Taken from the Army Energy Program website at http://army-energy.hqda.pentagon.mil accessed 22 April 2013
[2] Taken from the Operational Energy Strategy dated 1 March 2011 and accessed 25 February 2013 at http://energy.defense.gov/OES_report_to_congress.pdf

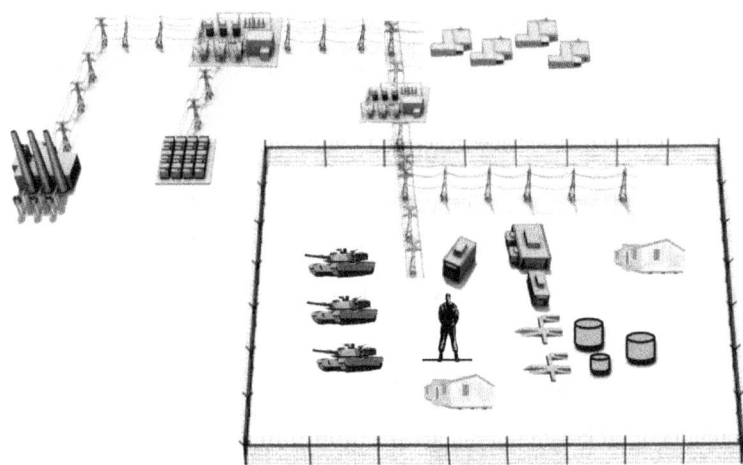

Figure 1.1 An installation that is dependent on external sources of power

Physical and cyber attacks to the energy grid and supporting elements is a long-known risk. Open sources report has disclosed to public they have information of cyber attacks against power system controls from outside the U.S. Multi-city outages and rolling blackouts have become more common. Natural and man-made disasters along with disgruntled employers are the primary threat to the grid along with sophisticated enemies such as China. Our current nation power infrastructure is fragile and not resilient – in a time of war it would be an easy and crippling first target. We, the DoD, must be prepared to execute our national security mission. Without power this is not possible. Table 1.1 contains a threat matrix as a function of physical, cyber, and natural attacks. The table shows the consequence and the target. From that table you can see that 1) critical infrastructure is probably the most vulnerable to a wide range of threats, 2) disruption can be caused by a host of people, 3) little differences between natural and man-made, and 4) cyber attacks are the most persuasive.

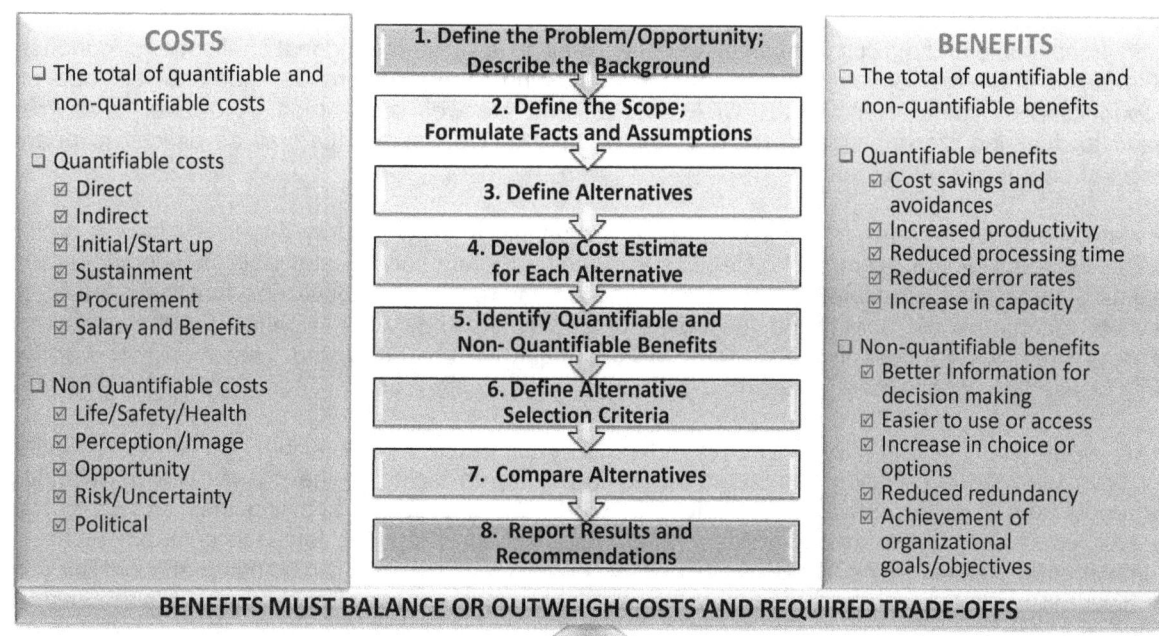

Figure 1.2 *The eight step CBA process (U.S. Army, 2012)*

When this guide refers to the Army enterprise, it means that initiatives should be evaluated based on the benefits they provide to the Army as a whole, not to any individual organization. A CBA makes the case for a project or proposal, weighing the total expected costs against the total expected benefits, over the near, far, and lifecycle timeframes, from an Army enterprise perspective.

Energy security is particularly challenging. Like aesthetics it can often be hard to quantify. Many of the benefits must be expressed in non-economic terms and are difficult to quantify. However, it is a component of ability to project forces and conduct the mission of the armed forces.

1.3 Documenting a CBA
The preferred method of documenting a CBA is through the use of narrative document such as a report. In general, a narrative description better details the situation and analysis that are necessary for a CBA. In Appendix A, this guide includes an example CBA focused on energy security.

The final CBA presented to the decision maker must provide a recommendation that meets the objective of the CBA, as well as a value proposition that supports the recommendation. A value proposition is a clear statement that the benefits more than justify the costs, risks, and tradeoffs/billpayers. In other words, a value proposition is a short statement that describes the tangible results/value a decision maker can expect from implementing the recommended course of action and its benefit to the Army. A value proposition should tell the decision maker exactly what can be achieved by implementing the recommended course of action.

An example of a strong value proposition is: "By burying power lines from Fort YYY installation's main transformer to the power distribution substation at XXX Airbase, the amount of time the airbase and supporting facilities will be without electric on average will be reduced from .6% (52 hours) to .3% (27 hours) annually." It is specific, and reports tangible, attractive results. An example of a weak value proposition is: "By burying power lines from Fort YYY installation's main transformer to the power distribution substation at XXX Airbase energy security will be increased."

1.4 Literature Review

Identify quantifiable and difficult to quantify benefits (Step 5) is by far the biggest challenge for conducting CBA analysis of energy security projects. Basic engineering economics and decision analysis tools exist for conducting the other steps in the CBA process and are well understood. However, research is needed to describe the quantifiable (both economic and non economic) as well as difficult to quantify aspects of energy security projects.

The literature contains many references on how to quantify energy security especially from a national strategic perspective. For example the Nautilus Institute for Security and Sustainable Development (1998) presents a comprehensive review of the state of the art and makes the conclusion that there are no real analytical techniques for quantifying the value of energy security. Also, Hughes (2009) presents an Analytical Hierarchy Process very similar to the technique we are proposing. However, that work was mainly focused on national energy security.

The US Army Corps of Engineers came up with an analysis of the energy security on Army installations. They discussed the information mentioned earlier in our report including the possibilities of "islanding". This report then lists the different alternatives that they will use before applying their criteria to them. However, when they get to this point, each alternative has three options for each criterion. Either it is "good" with that criteria, "no go", or is somewhere in the middle. This does not take into account the range that these alternatives can fall on. However, that they used a large number of alternatives is the best idea to come out of this report.

The Department of Energy (2006) has developed an energy security assessment guide. This guide is designed for federal installations to support:

- Initiation of the energy security assessment process,
- Vulnerability assessment,
- Energy preparedness and operations planning,
- Remedial action plans, and
- Management and implementation.

This report gives a detailed process for conducting vulnerability assessment planning, illustrated in Figure 1.3. However, it does not contain any type of methodology for assessing technologies that mitigate these vulnerabilities. The report does address another interesting concept; the synergism between security and environmental issues.

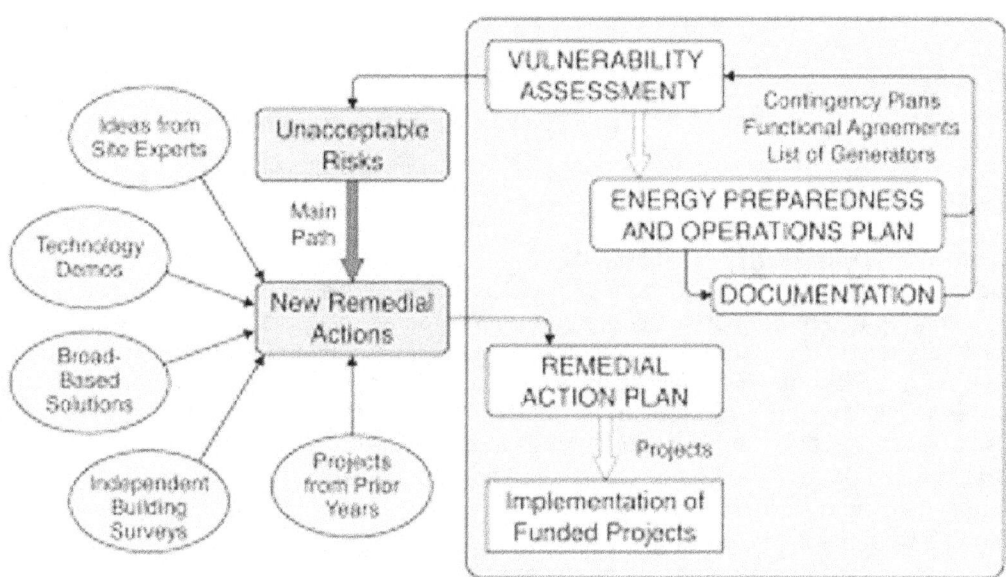

Figure 1.3 Energy security program flow diagram modified to show the transition from the
vulnerability assessment to remedial action[3]

James Lambert, in his report titled "Energy Security of Military and Industrial Systems: Multicriteria Analysis of Vulnerability to Emergent Conditions including Cyber Threats", chooses to distinguish *emergent conditions* (conditions that may develop and affect investment decisions in the future) as either evidence-based or based upon the "subjective advocacy positions of the various stakeholders" (Part II: Background). The general approach of the report focused more on deciding which portfolio of investments best matches a given future scenario best summarized as: "a set of scenarios comprised of emergent and future conditions that influence energy security" (Part III: Technical Approach).

Lambert states that the results from this report present the stakeholders with a set of high performing alternatives, as well as a small set of scenarios that need to be more carefully studied. The difference that makes itself apparent between our report and theirs is that we are trying to maximize desirable achievement with one energy portfolio as dictated by the stakeholders. For instance, if we consider cyber threats to become more significant in the future, then we can increase our share in a portfolio that is more secure against them by increasing the weight of resistance to them. We have explored different future scenarios by generating alternative weights to use if the stakeholder is more concerned with achieving Netzero or energy security.

Alsfelder, et al (2012) used a multi objective decision analysis (MODA) and data envelopment analysis (DEA) to evaluate value as a function of life cycle costs of different energy security measures and renewable energy for military installations. This portfolio approach in an effort to capture both the true costs and to develop technology feasible alternatives. This study will propose methods, processes, and tools for the decision makers to compare the portfolios' value and cost effectiveness. This paper will culminated in a demonstration of the methodology to illustrate its' utility using Fort Carson. Unfortunately, this methodology was developed for integrated portfolios of projects and did not work well for single projects.

---

[3] Taken from Department of Energy, Federal Energy Management Program, Performing Energy Security Assessments —
A How-To Guide for Federal Facility Managers, accessed at http://www1.eere.energy.gov/femp/pdfs/energy_security_guide.pdf,
accessed December 15, 2011

Chapter 2
Step 1 in the CBA
Define the Problem/Opportunity and
Describe the Background

2.1 Problem or Opportunity Statement

The first and one of the most important steps of the CBA process (see Figure 1.2) is to define the initiative or proposal using a problem or opportunity statement. A problem statement clearly defines the problem, mission need, and required capability. An opportunity statement is similar to a problem statement, but is focused on taking advantage of a favorable situation. When developing a problem or opportunity statement, the key is to state the problem or opportunity in terms of the organization's mission that requires a solution to describe what the effort intends to accomplish.

- What required performance or outcome is not being achieved?
- What is the perceived vulnerability gap?
- Who and what are impacted by this problem?
- Specifically, who are the customers or stakeholders?
- Briefly describe the process for providing the procedure, product or service where the problem or improvement opportunity occurs and how and why it occurs.

Example of a weak problem statement: "By installing solar cells we can improve energy security." This statement is vague, does not identify the problem, and does not propose a solution to the problem.

An example of a good problem statement with respect to energy security might read as *"For the last five years, on average Fort XXX has lost power for 60.3 hours (32% was during normal working hours) for greater than 50% of the base due to downed power lines on post. This has affected quality of life and the ability of the XX Division to conduct training missions and in a time of war could preclude power projection."* Note that the problem statement does not propose a solution.

The problem/opportunity statement should also be defined using clear, results-oriented language and be unbiased as to a recommend solution. The more precisely the problem/opportunity can be defined, the greater the likelihood that the analysis will meet the needs of the decision maker.

2.2 Objective

The objective statement describes the purpose of the CBA: what is the decision to be made, and how does the CBA inform and support it? What is the purpose of the analysis?

Examples of objectives that may be appropriate:
- To inform Congressional decision on funding for energy security projects at Fort Benning.
- To inform senior leader decision on the use of renewable energy and how that relates to energy security at Army installations.

The CBA preparer should identify key stakeholders early in the CBA development process. The decision-maker is one of the most important, if not the most important, stakeholder. They are the ones who best define the problem/opportunity and determine if the CBA is solving the right problem (or capitalizing on the right opportunity). The stakeholder's opinion in terms of what is important to them is of critical importance to the outcome of the CBA. The decision-maker must have an understanding of how to use the CBA once it is complete and how it will be implemented. The decision-maker does not need to understand the detailed analysis techniques used in the CBA, but should feel comfortable with the conclusions offered. The term "Voice of the Stakeholder" (VOS) is a phrase that is often used to describe their needs and desires. The VOS is an important input for developing the selection criteria (also known as evaluation / decision criteria) and identifying the benefits that will result from solving the problem (or opportunity).   Besides providing guidance to the CBA preparer, the stakeholders also help determine/validate the criteria which will be used to evaluate and compare CBA COAs.   Decision criteria are an outcome of the "Voice of the Stakeholder".   Needs and requirements of the Stakeholders should be translated into the means of evaluating COAs.

Stakeholders or customers are the functional process owners or the end users of the products and/or services flowing from the CBA.   In other words, it is any person or organization who will be directly affected by the outcome of the CBA.   They are the audience.   While there may be many stakeholders, the decision-maker(s) are usually the most important.   The analyst responsible for preparing a CBA should make every effort to identify the primary (most affected) stakeholders in order for them to be consulted through the CBA building process.   This helps to ensure that the CBA is meeting their needs and requirements (which will be covered later in this step).   This is done by soliciting their input at key points.

2.3 Background
The background and circumstances define and assess the current state/condition.   It provides the contextual information needed to fully understand the problem, need, or opportunity to be addressed in the CBA.   Defining the current state is the method of identifying system characteristics (current process or state of operations), users, and stakeholders, as well as the problems with the current system.   The information should be detailed to a level where all stakeholders can understand and support conclusions drawn from the analysis.   When the creator of the CBA neglects to spend time on the background and circumstances of the situation, stakeholders are given no understanding of the problem or why alternatives are being proposed.   Background information must be incorporated into all areas of the introduction to the CBA: problem statement, objectives, scope, assumptions, and constraints.

Chapter 3
Step 2 in the CBA Process
Define Scope, Formulate Facts, and Assumptions

3.1 Scope

The scope of the analysis defines the range of coverage encompassed by the project along specific dimensions such as time, location, organization, technology or function. The CBA should state the involved stakeholders, period of time that the analysis covers, as well as organizations or requirements not covered or addressed in the analysis. Defining the scope of the CBA is critical because it keeps the CBA focused on the things that matter. A well-scoped CBA should reinforce the problem statement defined in Step 1.

3.2 Formulate Facts and Assumptions

A fact is something that is empirically true and can be supported by evidence. Include only *relevant* facts – those items of information that have a direct bearing on the CBA being developed. Facts can include constraints, or limits placed on resources for the project. These may include: organizational policies or procedures, funding considerations, physical limitations, and/or time-related considerations. These policies/considerations could stem from technical, environmental, ethical, or political constraints. External constraints or barriers are normally beyond the control of the analyst and provide limitations within which analysis takes place. While constraints are usually beyond the control of the analyst, they are not necessarily beyond the control of the organization.

Assumptions identify conditions that must exist or events that must occur in order for the recommended COA to be successfully implemented. An assumption involves a degree of uncertainty. Assumptions play a critical role in explaining CBA results, in building credibility for the case, and in reducing and measuring uncertainty in projections. For this reason, regardless of the impact on the analysis, identify all pertinent assumptions. Do not confuse assumptions with facts or statements that, with research, could be presented as factual data.

Here are two examples of assumptions:
- One example might be that "an installation is considering installing photovoltaic cells on all carports and that these arrays can be retrofitted into existing structures." In this particular instance, however, there may have been no reason why this assumption could not be verified with research and presented as a fact.
- If the organization is considering a solution that would require a change to a federal law, the analysis might include an assumption that any required legislative changes would be approved by higher headquarters and enacted by Congress. This is something that is clearly beyond the local organization's ability to control or to know for certain.

In order to properly constrain the analysis, facts and assumptions should be established and fully documented early in the process. This is done to preclude a recommendation that is not feasible or cannot be implemented due to factors beyond the control of the implementing organization. An alternative is feasible only when it satisfies all the restrictions. Facts and assumptions should discuss anything that could impact or affect the quality of the cost estimate as well as be used to highlight cost issues of importance to decision-makers.

Chapter 4
Step 3 in the CBA Process
Define Alternatives

4.1 Introduction

As was mentioned earlier in this guide, one of the most important goals of the CBA is to prepare unbiased solutions or recommendations for the decision maker, based on critical reasoning and reliable information (data). Alternatives can be intuitively obvious to the analyst or team preparing the CBA or they may take a determined effort to define. There is no magic to coming up with a sufficient number of alternatives. Creativity is key to developing effective solutions. Often, groups can be far more creative than individuals. However, those working on solutions should have some knowledge of or background in the problem area.

4.2 Define the Status Quo

Functionally, it is the existing operational capability of the program on the start date. It also takes into account the future plan of the organization, such as planned and scheduled changes and/or enhancements to the existing program and should reflect a review of mission and strategic goals. Generally, the only time that a status quo does not exist is when a solution is being proposed to address a new requirement or mission.

Not all situations requiring a CBA will include the status quo as a viable alternative. If the status quo does not conform to the mission and strategic goals, or does not capably address the requirements or objectives, then it should not be considered as an alternative. Also, higher leadership might direct against considering the status quo as an alternative, and recommend development of COAs in a different direction. A CBA that does not include the status quo as a COA must be fully justified to the organizations reviewing the documentation.

The status quo is often used as a baseline for estimating cost, savings, cost avoidance, and other aspects of how a given COA represents improvement over the baseline. As a COA, the status quo serves to highlight any issues, defects, shortfalls, or strengths inherent in the current state. We compare all COAs, to include the status quo in Step 7. The decision maker can use this information to determine what choices need to be made or how to capitalize on the current situation. For example, if higher efficiency in delivering products to command posts is required, and the status quo shows that there are far too few vehicles to meet the new requirements, then alternatives can be drafted addressing the need for more vehicles.

In order to be used as a "measuring stick" the costs and benefits of the status quo must be fully documented and included in the analysis. If the status quo is not included in analysis, a thorough explanation is necessary. Without the status quo costs it is very difficult to evaluate the benefits associated with the new program. Where a status quo exists, omitting it from the cost benefit analysis will reflect negatively upon the analysis and the credibility of realizing any proposed quantifiable benefits. If the status quo includes scheduled/planned/directed changes or enhancements, these should be included in the estimation/documentation. However, the analyst must be careful when considering factors that may change in a few years. The cost of operating the status quo until the new system or project is fully operational (known as parallel operations) will be a part of the cost of all other alternatives in the cost-benefit analysis. These costs are referred to as Phase-out or Transition costs.

4.3 Define Alternatives / Courses of Action (COA)

The CBA alternatives (or COA's) should reflect a review of the mission and strategic goals and should address the base requirement as outlined in the problem statement. The status quo alternative is always the first alternative. All alternatives should be viable solutions to the problem statement. Avoid using a

COA that is clearly not a reasonable solution. It is better to have fewer viable alternatives than many weak ones.

The CBA preparer should use screening criteria to ensure solutions being considered can solve the problem. Screening criteria defines the limits of an acceptable solution. As such, they are tools to establish the baseline products for analysis. A solution may be rejected based solely on the application of screening criteria. Five categories of screening criteria are commonly applied to test a possible solution:
- Suitability - solves the problem and is legal and ethical. The COA can accomplish the mission within the decision - maker's intent and guidance.
- Feasibility - fits within available resources.
- Acceptability - worth the cost or risk.
- Distinguishability - differs significantly from other solutions.
- Completeness - contains the critical aspects of solving the problem from start to finish.

The number of alternatives should be controlled by avoiding similar but slightly different alternatives (variations on a theme) and by early elimination of non-viable alternatives. The reasons for eliminating potential alternatives should be included in the CBA documentation. Some of the criteria used as a basis for eliminating non-viable alternatives are:
- Unacceptably high cost/performance,
- Lack of compliance with established constraints,
- Dependence on assumptions that are unrealistic,
- Inability to meet Initial Operation Capability (IOC) or full operational capability (FOC) requirements, and
- Political considerations such as environment, world opinion, treaty compliance, etc.

Because each project requiring a cost benefit analysis is different, the following questions should be considered as guidelines during the preparation, review, and validation of CBA alternatives:
- Do the alternatives reflect a review of mission and strategic goals? Have all feasible alternatives been considered? Are all alternatives presented feasible?
- Are the alternatives distinctly different?
- Have the alternatives that were eliminated from the analysis been clearly identified and has a rationale been provided for their elimination?
- If other Government organizations can provide the desired product or service, have they been identified as alternatives?
- Are tradeoffs of each alternative clearly stated? Unavoidable and difficult tradeoffs should not be hidden.

4.4 Describe Second and Third Order Effects (Cause and Effect)
As part of a through discussion of each COA/ alternative, an analyst should also pay careful consideration to the "effects" that any COA may cause if implemented.

In addition to the primary intended result or consequence of a decision, there can be second- and third-order effects. The concept of second- and third-order effects is based on a sequential cause and effect relationship. When a decision is made, it is the cause of effects A, B, and C. Each of these effects can in turn become the cause of other effects, and so on as the full impact of the decision is felt. Ensure to analyze an alternative in terms of its second- and third-order effects as shown in Figure 4.1 in order to capture the interdependencies as shown in Figure 4.2. To identify second and third-order effects, the analyst should ask questions such as: "If this action is implemented, what will happen? And what will happen as a result of that?" Because decisions have consequences, analysts must understand what those consequences are and assess their impacts not only within their immediate organization, but horizontally and vertically within the larger organization (Army-wide) as well. Finally, one of the most important questions is: "If a recommendation is adopted, will it create a bill for another organization?"

Again, if it creates a bill for another organization, the analysis/assumptions should be vetted with that organization.

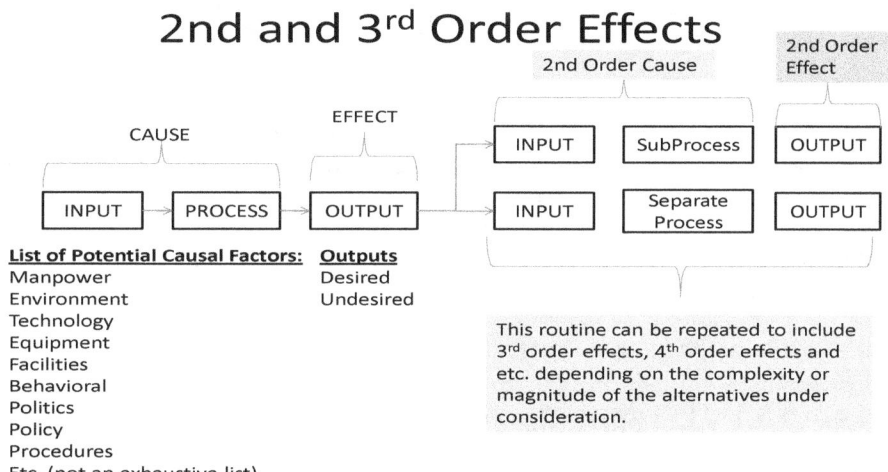

Figure 4.1 *Second and third order effects*

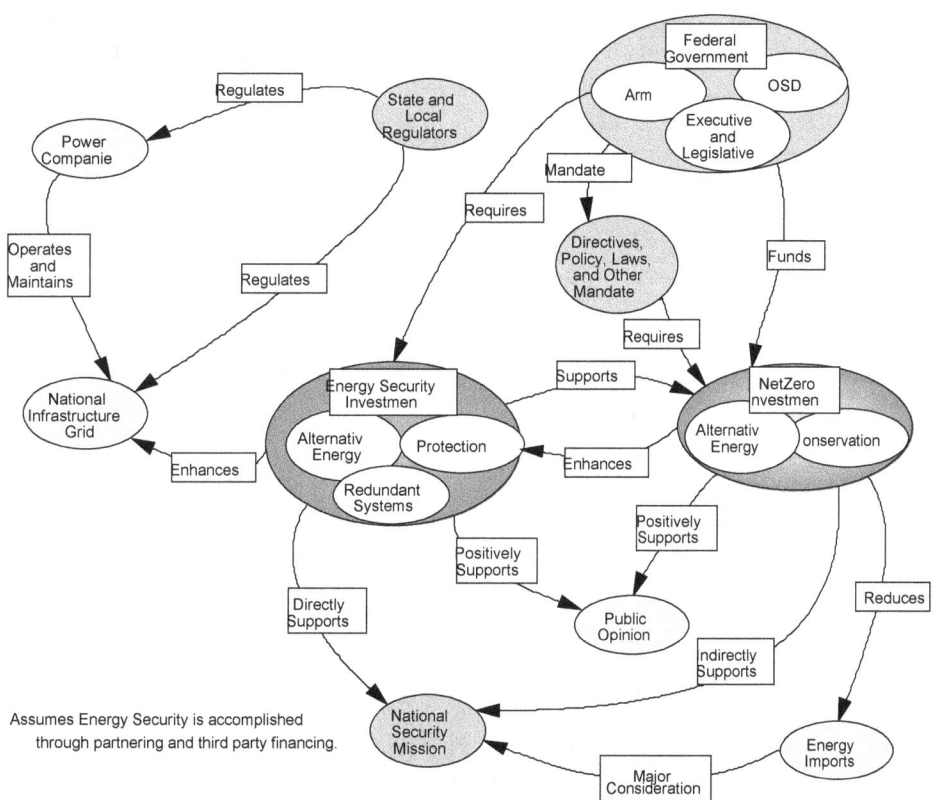

Figure 4.2 *Interdependencies of energy security and Netzero*

Chapter 5
Step 4 in the CBA Process
Develop Cost Estimates for Each Alternative

5.1 Cost Concepts

Cost analysis is a critical element in the CBA process. Cost estimates support management decisions by translating resource requirements (e.g., equipment and personnel) associated with programs, projects, or processes, into dollar values. It is one of the most challenging steps in the CBA process. Using the best data available will result in the best estimate. Much of the analyst's time will be spent on obtaining data. Finally, it is important to capture all the costs related to the initiative or project for which the CBA is being developed.

Costs can be categorized as direct or indirect. They also can be categorized as fixed or variable and as recurring or non-recurring.

- A fixed cost is a cost that remains the same regardless of change in output, while a variable cost is one that changes with changes in output.
- A recurring cost is one that is incurred repeatedly for each organization and/or product/service. This cost must be programmed and resourced each year.
- A non-recurring cost is a cost that will happen only once.

Cost estimating is an iterative process that may require reevaluating previous steps, with a systematic approach to develop accurate and timely estimates. Figure 5.1 presents the steps in the cost analysis process.

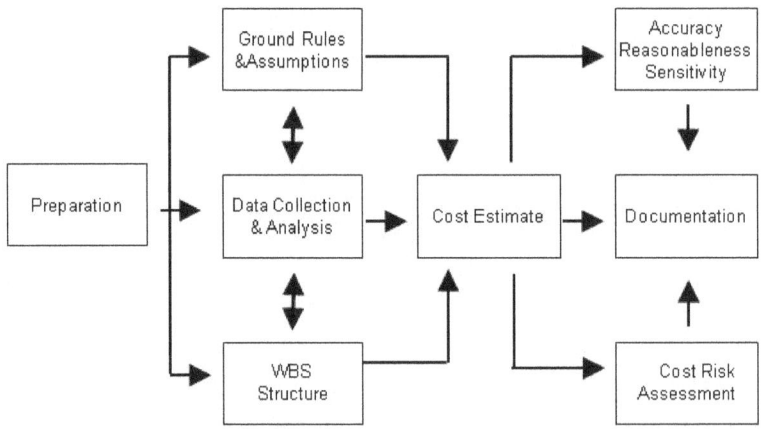

Figure 5.1 *Cost estimating process*

Once the purpose is understood, it is important to agree on the end product (deliverable) that is going to the customer. This is also the time to ensure that the scope of the cost estimate is understood and defined and the level of detail necessary is adequate to support the alternatives under consideration. Finally, the analyst should understand what the time constraints are that he/she will work under in preparing the CBA. The more cost detail required, the more time and staff the estimate will require.

Data is a critical component of the cost estimate. Data quality affects the estimate's overall credibility. This step includes the process of identifying, collecting, and analyzing data before applying cost estimating tools within the analysis process. Data collection can be a time consuming process and continues throughout the cost estimate. In general, data can be associated with activities that generate costs; activities that are defined or described using schedules or dates; and technical requirements of equipment and material.

Develop and implement a formal data collection plan. Data collection entails the following tasks:
- Identify the types of data available (e.g. cost, programmatic, schedule, technical).
- Collect cost data with supporting documentation.
- Determine which estimating methods, tools, and models will be used with which data sets.
- Verify, validate, and adjust (normalize) the data. Cost data is adjusted in a process called normalization, which improves the quality of the data. In short, normalization ensures apples to apples comparison vs. apples to oranges.
- Collect data continuously throughout the pre-cost estimating process.

The cost estimate must be documented detailing the source of all data and the processes used to analyze the data. Documentation should provide enough detail for another person to track the cost-estimating process from definition to conclusion, enabling an analyst (even unfamiliar with the original analysis), to modify the analysis at a later date. Unsubstantiated cost data casts uncertainty over the entire CBA process. Where possible manufacturer's data, other similar results, high-resolution models, etc., should be used to predict and defend all cost estimates and other derived benefits.

Cost estimates predict future events and thus by nature have risk and uncertainty. The key is to identify the risk so it can be managed and controlled. There are many tools and techniques, such as probability theory, game theory, Monte Carlo technique, Delphi technique, and decision trees to aid in making quantified risk assessments. Risk analysis examines the likelihood that actual results will fall within a specified range around a predicted point estimate, using probability concepts. Once the analysis is complete, the risk must be explicitly defined for the decision maker. Every LCC estimate should have a risk analysis.

Finally, costs should be analyzed and organized with respect to their occurrence. That is, some costs are onetime costs (non-recurring) that only arise once in and others cost are recurring (costs are generated each time an item is produced or service performed) such as operations and maintenance costs.

Trade-offs (or opportunity cost) describe the situation where resources are limited, requiring the pursuit of one action over another. The opportunity cost of an item is what you give up to obtain that item. The opportunity cost of any action is simply the next best alternative to that action - or put more simply, "What you would have done if you didn't make the choice that you did". Incorporating a discussion of trade-offs is an important consideration of cost analysis. Each of the alternatives in a CBA should be evaluated in terms of what must be given up in order to be pursued. Identifying trade-offs is conducted by evaluating each COA individually and not by comparing one COA against another. That is, examine each COA in isolation.

Tradeoffs can be described in financial and non-financial terms such as describing an activity to carry out which precludes doing something else. Where feasible, the analyst should attempt to not only describe the tradeoffs but also quantify them. For example, there are numerous laws, executive orders, DoD directives dealing with alternative energy tradeoffs within these constraints is important.

Chapter 6
Step 5 of the CBA Process
Identify Quantifiable and Difficult to Quantify Benefits

6.1 Benefits Analysis Overview

Benefits of a chosen alternative are results expected in return for costs incurred. They are the quantitative and qualitative results expected or resulting from the implementation of a project/initiative (which may include but are not limited to the following: equipment, facilities, hardware, systems, etc.). The following definitions or measurements describe benefits: effectiveness, physical yield, products, morale, quality of life, and timeliness.

When preparing a CBA, identify all benefits. Benefits justify the costs identified in the CBA. Identify both financial benefits (i.e., those measured in dollars) and non-financial or functional benefits. Both are essential to the analysis and selection of a preferred COA. Each benefit must be clearly and distinctly identifiable, and should not duplicate any other measure.

Quantifiable benefits have numeric values such as dollars, physical count of tangible items, or percentage change.

Financial benefits are always quantifiable and are measured in dollars:
- *Cost reduction.* A reduction in the number of dollars needed to meet a customer-established requirement by improving a process or function.
- *Savings.* A cost reduction that enables a manager to reallocate funds within the budget or program period.
- *Cost avoidance.* Any cost reduction that is not saving.
- *Revenue generation.* An increase in the dollars that flow into the Army, over and above appropriated funds, or over and above the expected amount of customer funding received through a revolving fund.
- *Productivity improvements.* A reduction in personnel time and effort requirements associated with a function or assigned task. In most cases, a productivity improvement will also result in a savings or cost avoidance.

Table 6.1 lists some examples of quantifiable economic benefits. This list is not all inclusive, nor is it intended to provide precise definitions of the benefits listed. It is only meant to be illustrative of benefits categories that could be applicable to program objectives.

Table 6.1 Metrics for assessing the value of an energy security project

Quantifiable - Economic			
Metric	*Discussion*	*Assessment Means*	*Comments*
	Non Reoccurring Costs		Should present this as both an Equivalent annual cost (EAC) and/or net present value (NPV) based upon total ownership cost (TOC)
Installation Costs	Should include all training, construction, support infrastructure, etc.	EAC and/or NPV based upon TOC estimate	
	Reoccurring Costs		Should present this as both an Equivalent annual cost (EAC) and/or net present value (NPV) based upon total ownership cost (TOC)
O&M Costs	Operations and maintenance (O&M) costs	EAC and/or NPV based upon TOC estimate to include training, people, supporting infrastructure, consumables, etc.	Increases in O&M costs should also be captured. Many energy security projects increase base operations and facilities management costs.
Energy Consumption	Energy costs can increase for some projects such as generators	EAC and/or NPV based upon TOC estimate	Systems designed to produce redundancy will in most cases cause high annual energy costs.

Examples of other, non-financial, quantifiable benefits and methods of measurement include but are not limited to:

- Number of commodities or items produced for each alternative (such as the number of meals served, hours flown, or components manufactured).
- Flight hours per month or number of trucks serviced per year.
- System reliability in terms of probable failure ratio Maintainability/supportability measures (such as mean-time-to-repair or average downtime).
- Accuracy, timeliness, and completeness of data produced by a system, performance and operational effectiveness.

Difficult to quantify benefits do not lend themselves to direct and quantitative measures. Although subjective in nature, qualitative statements can make a positive contribution to the analysis. The CBA preparer should use the best analytical practices in order to include difficult to quantify benefits in the analysis. Some examples of difficult to quantify benefits are morale, compatibility, quality and security, and readiness. Generally speaking, difficult to quantify benefits do not provide as much support for a COA as quantitative benefits do.

6.2 Identify, Estimate, and Evaluate Benefits

A CBA must include all significant benefits (quantifiable or difficult to quantify) in the benefit analysis portion with supporting analytics. Difficult to quantify benefits should be described in narrative form. Be sure to validate and coordinate all the benefits by the functional proponent (or the organization responsible for the basic requirement) and appropriate activities. It is strongly recommended that identification and documentation of benefits begin early in the evaluation process.

The following steps are recommended to identify benefits and establish quantitative measures for benefits where possible:

- Identify all resources flowing into the project and the resulting outputs and outcomes flowing out of the project.
- Determine and list the benefits of each alternative, both quantifiable and difficult to quantify.

- Define each benefit in relation to the alternatives in the CBA. All benefits included must be relevant to the analysis. Each benefit must be clearly and distinctly identifiable from all other benefits; it should not duplicate or overlap any other measure.
- Develop a quantitative measure for each benefit where possible. This will allow direct comparison of alternatives for each benefit.

The benefit estimating process is similar to cost estimating (discussed in Step 4.) Data must be collected from appropriate sources and analyzed. Relationships among data must be identified. The economic life (the period during which the alternative provides benefits) of the alternatives and the fiscal years (FY's) when benefits accrue must be carefully considered. Some benefits may not be accrued until later in the economic life of the alternative.

During the quantifying analysis process, assumptions and judgments will influence the results. The analyst may have to make value judgments. They should inform the decision-makers of how the benefits were identified and measured. The analyst must avoid double counting of any identified quantifiable benefits, which will lead to skewed estimates of benefits.

6.3 Evaluating Difficult to Quantify Benefits
The following are techniques for evaluating difficult to quantify benefits:
- Enumeration is a simple listing of the difficult to quantify benefits associated with each alternative for comparison purposes.
- Rank difficult to quantify benefits by their relative importance to the goals and objectives. Such a ranking describes the degree to which each alternative achieves a given objective. The ranking provides a description of all benefits and how each contributes to the project's goals; it explicitly identifies the differences among alternatives. An example would be the quality of a report prepared automatically or manually. The judgment of which alternative yields the best quality report would assist in the overall ranking of alternatives. In addition to relative ranking, weights may be assigned to each benefit, so that a point total may be calculated for each alternative. Even if numeric scores are calculated, this analysis is by nature very subjective; it requires a consensus on the relative importance of the benefits.

Difficult to quantify benefits, in most cases, can become quantifiable with an appropriate measuring/counting methodology. For example, morale is often described in difficult to quantify terms such as good, bad, or something else. A survey or other measuring/counting methodology can be designed and used to measure the level of morale in more quantifiable terms like a value of 1 could equal a bad morale, 2 could be assigned to good morale and 3 to excellent morale and etc. Quantifying difficult to quantify benefits facilitates making meaningful comparisons of the benefits.

It is crucial to develop means of quantifying the value of the following characteristics: level of energy security, compliance with laws and mandates, renewable energy use, environmental impacts, and social/public perception. However, their value cannot be calculated economically. Therefore, a system for each metric has been developed in order to assign them value.

In our methodology, energy security can be calculated from a value matrix on a scale of one to one hundred by comparing energy security risk (impact) versus disruption in services (duration). The military post is broken down into a system, an installation, and the mission. Each of these levels could incur a security breach for either a temporary or long-term/permanent duration. The impact that the energy security project being considered is ranked as low, intermediate, or high. By organizing the matrix as shown in Table 6.2 in this manor, a solution for a single system that has a low impact earns a value of 10. Likewise, a project with a high, long term impact, the overall mission of a military unit earns a value of 10.

Energy Security Risk

Project Impact	System		Installation		Mission	
	Temporary	Long Term/ Permanent	Temporary	Long Term/ Permanent	Temporary	Long Term/ Permanent
High	40	50	70	80	90	100
Medium	20	40	60	70	80	90
Low	10	20	30	40	60	80

Duration

Table 6.2 *Energy security risk versus disruption in services duration matrix*

One way to think of the consequences to an installation of an investment in energy security assign values from Table 6.2 accordingly. For example
- Low Impact. Minor contribution to improved security of system, installation, or mission.
- Medium Impact. Moderate contribution to improved security of system, installation, or mission.
- High Impact. Significant contribution to improved security of system, installation, or mission.

Note that every installation is different. For example, consider family housing. If the housing at Fort Stewart were without power this would probably be considered a installation issue and a score assigned accordingly. Whereas, disruption of power to housing at Fort Wainwright would probably constitute a mission issue.

Referring to Table 6.2, installing a generator to provide back-up power for a single building would be classified as a system level project that has a medium, temporary, long-term impact; thus, receiving a score of 20. Burying power lines on a military post would be classified as an installation level project with medium, long-term impact; thus, receiving a score of 70. Installing a renewable energy based micro grid on a military post would be classified as a mission level project that has high, long-term impact; thus a score of 100 might be warranted.

In regards to energy security at a military post, a system is considered to be any device (i.e., component level up to a building or group of buildings) that aids in power production/distribution or network security. An installation is a network of systems, or in other words, the actual post. The mission is based on the goals of the people working inside the installation. For example, West Point is a military post that is also considered an installation. The mission of West Point is to provide the Academy with the best environment possible for producing leaders in America's Army. The backup generators for the cadet barracks at the United States Military Academy are a system within the installation.

Table 6.3 contains a summary of the various metrics used to assess non-economic or difficult to quantify benefits for energy security. We use the same system, installation, and mission categories. Like energy security, the values range between zero and one hundred.

Table 6.3 Non economic benefits

Quantifiable – Non Economic		
Metric	*Discussion*	*Assessment Means*
	Security	
Security	See Table 6.2 with a max value 100	See the assessment matrix in Table 6.2
	Other Tangible But Secondary Benefits	
Renewable Energy	Score it between 0 and 100	See Table 6.3
Energy Reduction	Score it between 0 and 100	See Table 6.4
Environmental Impacts	Score it between 0 and 100	See Table 6.5
Social/Public Perception	Because energy security in many cases also has a NetZero component they can have a positive public perception	See Table 6.6

For renewable energy, energy reduction, environmental impacts, and social/public perception the challenge is how to define "low," "medium," and "high". Every installation is different. Whereas, renewable energy[4] is viable at Fort Irwin it would be a challenge at Fort Drum. This is certainly achievable at some installations and not cost effective at others. Table 6.4 provides a means for quantifying "low," "medium," and "high" for renewable energy. Note that renewable energy can be categorized into two separate categories for our work: 1) projects such as solar panels that return energy to the grid or 2) projects that can be taken off the grid and can contribute to energy security at the system/installation/mission level. Those projects that are truly renewable that contribute to energy security should be scored higher.

Table 6.4 Renewable energy scoring methodology

Impact/Level	System	Installation	Mission
High – Truly Security	40	80	100
High - Return to Grid	20	60	80
Medium - Truly Security	20	60	80
Medium - Return to Grid	10	30	50
Low/None	0	0	0

"Medium" and "High" are truly qualitative terms in this context. However, for renewable energy we offer the following guidance. We define "High" as meeting the objective energy needs at the system, installation, or mission level. A graphic showing objective energy is present in Figure 6.1. "Medium" can be defined as making a major contribution to meeting the objective energy needs.

---

[4] The Nation Defense Authorization Act of 2007 requires 25% of an installations power be renewable by 2025. Also, the Energy Policy Act of 2005 established green power purchasing goals for the federal government, whereby the 7.5% of electricity used by federal agencies must be obtained from renewable sources by 2013, 3% in FY07-09, and 5% in FY10-12. Executive Order 13423 now requires at least half (50%) of the required renewable energy consumed by an agency in a fiscal year to come from sources placed in service in 1999 or later and to the extent possible, the agency implements renewable power generation projects on agency property for agency use.

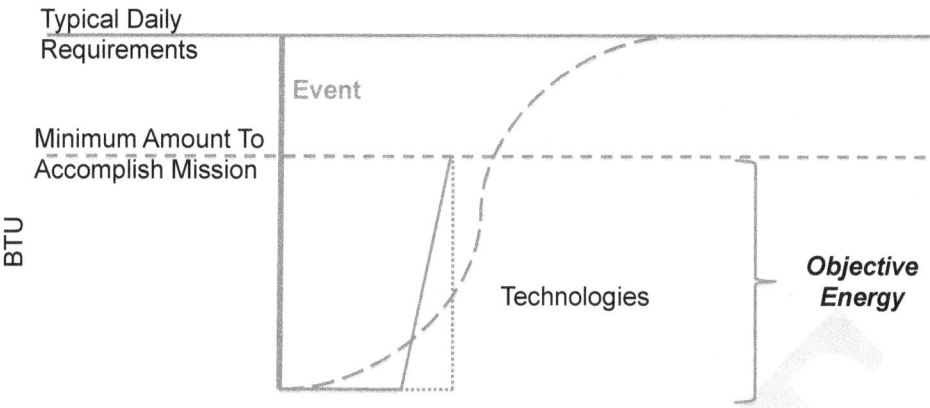

Figure 6.1 Graphic showing operational energy

Energy reduction is a critical component of any energy security. Table 6.5 captures the scores for energy reduction. Note that 0 values were assigned for both neutral and increased energy consumption

Table 6.5 Energy reduction scoring methodology

Impact/Level	System	Installation	Mission
Decrease Energy	20	60	100
Neutral	0	0	0
Increase Energy	0	0	0

Environmental impacts is an all encompassing term and can include carbon and solid waste reductions (or increases). Using the same methodology as prescribed in Table 6.5, we can qualitatively assess environmental impacts as presented in Table 6.6.

Table 6.6 Environmental impacts scoring methodology

Impact/Level	System	Installation	Mission
Decrease Impacts	20	60	100
Neutral	0	0	0
Increase Impacts	0	0	0

Table 6.7 can also be used for social/public perception. With regards to energy, any positive increase in public perception is important. Projects at the installation/mission level would probably receive national media attention and thus warrant the higher scores.

Table 6.7 Public perception scoring methodology

Impact/Level	System	Installation	Mission
High	40	80	100
Medium	10	30	80
Low/None	0	0	0

6.4 Define Alternative Selection Criteria

Alternative selection criteria are the standards used to rank the alternatives in order of preference, and to make the decision. After collecting and analyzing data for the proposed alternatives, and completing cost estimates, the decision criteria for selecting the "preferred" alternative must be determined. In many cases, the total cost or primary benefits are part of the selection criteria. A CBA must contain documentation that defines decision criteria and their impact in making the recommendation of the preferred alternative. It is important to customize the criteria to the CBA. For example, if an organization wishes to buy a new passenger vehicle for its fleet, some of the criteria that would go into the evaluation of the alternatives could include size, mpg, number of seats, etc. Note that advanced analysis techniques such as data envelopment analysis (DEA) could be used for assessing the relative effectiveness of an alternative based upon multiple criteria.

Use criteria to compare alternatives accurately and consistently, to prioritize needs, and to document rationale of decision-making and thus increase transparency within the Army. Decision-makers use criteria such as value as shown in Figure 6.2 to examine the most important information and use it to evaluate the impact of the alternatives on the mission/objective. In addition to documentation that identifies the recommended decision criteria, every CBA must document the extent to which each alternative satisfies each of the decision criteria. Thus, the first requirement in this process is to develop a list of candidate selection criteria.

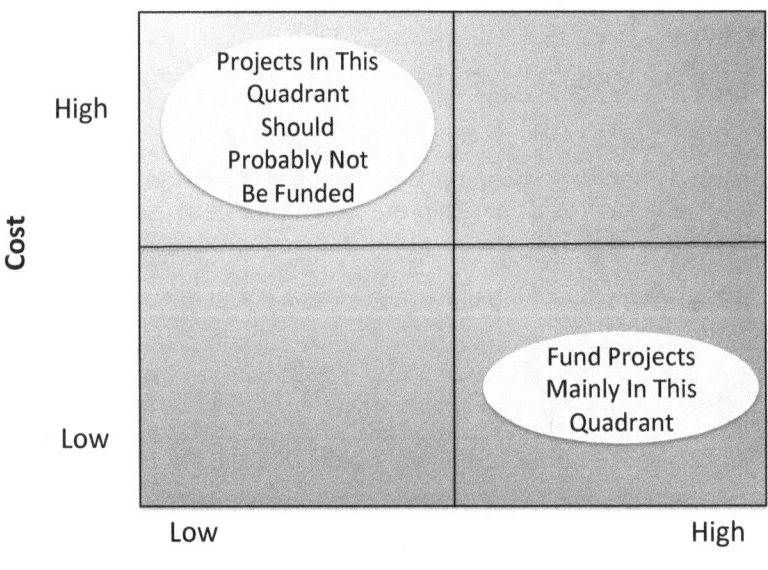

Figure 6.2 *Value versus cost project funding matrix*

All criteria will be highly tailored to the specific CBA, but there are characteristics that make selection criteria more legitimate and qualified to support recommendation of alternatives.

Selection criteria should:
- Be related to the alternatives and highlight differences between alternatives to support comparison,
- Selection criteria should reflect the costs and benefits listed in the analysis,
- Be unambiguous: the criteria must be clearly stated and the relationships transparent,
- Be concise and non-redundant,

- Provide a standard and consistency for comparison of alternatives,
- A means to expose all uncertainty, risk, and/or tradeoffs, and
- Inclusive of enough information to allow for an informed decision.

Steps in developing candidate selection criteria are:
- If possible, develop criteria prior to analyzing benefits,
- Identify relevant cost issues (See Step 4),
- Identify relevant benefits (See Step 5), and
- Identify negative impacts of each alternative course of action.

Next, pare the list of candidate criteria into a handful of the most meaningful factors that should be taken into account in selecting a course of action. This list will comprise the selection criteria against which each COA will be ranked, weighed, or judged.

A CBA preparer must consider the following questions, to ensure that all important points have been addressed.
- Are the selection criteria appropriately tailored to the problem statement/ requirement?
- Has appropriate consideration been given to both cost and non-cost criteria? If weighting of selection criteria has been used, has leadership agreed with the weighting?
- Do the selection criteria appear unrealistically biased to favor one alternative? (This is unacceptable.)

Chapter 7
Step 6 in the CBA Process
Define Alternative Selection Criteria

7.1 Introduction
This step is a continuation of the previous one. As was mentioned in that step, the analyst cannot successfully complete this step in the CBA development process without first defining the criteria that was the subject of Step 6 Define Alternative Selection Criteria. The analysis and calculations developed in Step 6 are critical to the tasks required in this step.

7.2 Compare Costs and Benefits
The essence of the CBA process is comparing the costs and benefits of two or more alternatives (including the status quo) in order to select the preferred alternative. As a general rule, the preferred alternative is the one that provides the greatest reward in relation to its cost. In situations where it is difficult to quantify benefits and measures of effectiveness, it is important to provide as much useful information as possible to support decision as to which alternative yields the most benefits.

Before an analyst can perform a comparison that will lead to a recommendation, there is one remaining area that must be discussed as it may be a consideration (criteria) for which a decision could be decided upon. It is risk analysis For example, the analyst or the decision maker may want to base the decision on the level of risk, preferring the COA with the lowest identified risk.

7.3 Risk Assessment and Mitigation
The CBA guide identifies risk assessment and mitigation as an important step in the process. A risk assessment is the identification and analysis of relevant risks associated with achieving installation energy security. Our Table 6.2 captures risk in terms of impact and duration. Most of our solutions should be mature technology.

7.4 Rank Ordering Projects
At this point, the analyst has carried out all the necessary analysis and should be ready to compare each COA with the intent of identifying a COA that best fulfills the objective/goal indentified in Step 1 of this guide. There are several tools / methods that an analyst can use to efficiently and effectively evaluate their analysis to determine the best COA to recommend. These tools / methods can utilize quantitative (financial) criteria, non-quantitative criteria, or some combination of both.

Multi-objective decision or value analysis (Kirkwood, 1997) uses an overall value function which combines the multiple evaluation measures into a single measure of the overall value of each evaluation alternative, or portfolio of projects. Thus, different mixes of projects in a portfolio may be compared to determine the appropriate mix for maximizing value. Multi-objective value analysis is useful for structuring the judgments used in assessing the value of projects that comprise a portfolio in an organization with multiple and conflicting objectives. Multi-objective value analysis methods are based upon structured objectives, evaluation measures, value functions, and weights.

A multiple criteria value function based upon weights and scores are used to rank alternatives. An additive value function is used for this research since it is common Keeney, 1992).

When weights have been determined for the current situation, the model can be used to find the right mix of projects to maximize value and support a combination of core outcomes within a fixed budget portfolio.

The mix of projects with the highest overall score adds the most value. We can then view projects as a function of cost or some other variable to make logical and defensible decisions.

When using multi-objective decision analysis or MODA, a structured approach must be taken to develop the weights and scores. In this paper we present weights and scores based upon the experience of the authors, literature, and input from subject matter experts. Stakeholder buy-in is critical with all parties agreeing to the framework. Sensitivity analysis can play a key role here to show how varying the weights over different ranges can have little or major impact on the objective function. Table 7.1 shows how a MODA type analysis could be used to evaluate three courses of action. Note that the weights were developed by authors and further stakeholder analysis is needed to refine the values.

Table 7.1 *Example of a decision matrix to evaluate non-financial selection criteria*

Criteria	Weight	COA 1			COA 2			COA 3		
		Data	Rating	Score	Data	Rating	Score	Data	Rating	Score
Operational Security	.60	Table 6.2	70	42	Table 6.2	60	36	Table 6.2	40	24
Renewable Energy	.1	Table 6.4	60	6	Table 6.4	30	3	Table 6.4	80	8
Reduced Energy	.1	Table 6.5	0	0	Table 6.5	0	0	Table 6.5	60	6
Environmental Impacts	.1	Table 6.6	0	0	Table 6.6	60	6	Table 6.6	0	0
Social/Public Perception	.1	Table 6.7	30	3	Table 6.7	0	0	Table 6.7	80	8
Total Score	1.00			51			45			46

The total scores in the bottom row would be compared with the cost of the COAs in order to arrive at a recommendation or decision. The criteria for the above decision matrix would come from the previous step (Step 6) of this guide. The criteria are user defined and should be coordinated with the decision maker to ensure that the criteria and their relative weighting satisfy his or her intent. It makes little sense to evaluate COAs using criteria that are of little importance to the person using the CBA to make a decision.

Chapter 8
Step 7 in the CBA Process
Compare Alternatives

8.1 Introduction

As noted above, financial and non-financial criteria must be combined in the same decision matrix. To demonstrate how financial and non-financial criteria should be compared in order to arrive at a decision or recommendation, we'll continue the above example. We'll add cost to the evaluation and display the financial and non-financial criteria in two separate tables as shown below. In Table 8.1, the COA 3 preferred project if non-financial data were used as the only selection criterion.

Table 8.1 Non financial considerations

Non-Financial Data		Rating		
Benefit Criteria	Weight	COA 1	COA 2	COA 3
Energy Security	.60	42	36	24
Renewable Energy	.1	6	3	8
Reduced Energy	.1	0	0	6
Environmental Impacts	.1	0	6	0
Social/Public Perception	.1	3	0	8
Total score		**51**	45	**46**

Financial Data	COA 1	COA 2	COA 3
Total Ownership Costs[5]	$200M	$190M	$240M

Another means to combine cost and benefit data in a single measure, a "cost-benefit index" or CBI can be developed as shown in Table 8.2. Using this measure, the preferred alternative is COA 1 with the lowest CBI.

Table 8.2 COA versus cost benefit data

Cost-Benefit Index (CBI)	COA 1	COA 2	COA 3
Cost	200	190	240
Benefit Score	51	45	46
Cost-Benefit Index	**3.9**	**4.2**	**5.2**

There is a variety of quantitative methods for project selection criteria that provide a definitive basis for ranking alternatives. Note that COA 2 is the cheapest in terms of cost. However, COA 1 has the highest value. Using CBS (costs divided by benefits), COA1 is the most effective option.

All cost estimates should include sensitivity analyses. It is not sufficient to present the decision maker with a single recommendation that is based on the 'most likely' costs, benefits, assumptions, and other factors. The decision maker needs to be informed about how well the alternative's rankings will hold up under reasonable changes to factors and assumptions. Describe how sensitive the recommendation is to changes. For example, a sensitivity analysis that addresses how sensitive the recommendation is to changes in cost might say, 'The cost estimate for this COA is $500K, but that estimate might prove to be

---

[5] Assumes that each of these projects will have the same life expectancy. Otherwise an equivalent uniform annual cost should be used.

incorrect. Analysis of this sensitivity has determined that as long as cost is $800K or lower, the recommendation would not change.' This gives the decision maker a 'comfort level' by assuring him/her that costs could vary considerably without changing the recommendation. On the other hand, if a recommendation is found to be extremely sensitive to small changes in cost, assumptions, or other factors, then a more in-depth analysis might be appropriate.

It is important to note that sensitivity analysis can address not only changes in cost and benefit, but changes in other factors as well, to include assumptions, constraints, scope, and weighting of selection criteria. A thorough sensitivity analysis should consider all these possible changes. It is recommended that sensitivity analysis be done especially on the most important selection criteria and the most important assumptions.

It may be helpful to divide analysis into two groups of factors:
- Those that are outside the control of an agency (i.e., assumptions) and,
- Those that an agency can influence or control to some degree.

Suggested steps for conducting a sensitivity analysis are:
- Choose several elements (costs, assumptions, benefits, etc) that appear to have the greatest impact on the results of the analysis and which are most subject to variance.
- Vary each one over a reasonable set of values while holding the other variables in the analysis constant relative to each other.
- Determine the impact of these changes on the net present value results and the ranking of alternatives.

Some factors that may warrant sensitivity analyses are:
- The effects of a shorter or longer economic life.
- The effects of variation in the estimated volume, mix, or pattern of workload; for example, the production rate or learning curve.
- The effects of potential changes in requirements resulting from either Congressional mandate or changes in functional responsibilities.
- The effects of potential changes in requirements resulting from changes in organizational responsibility at the site, installation, base, or Army command/direct reporting unit/Army service component command level.
- The effects of alternative assumptions on areas such as project operations, relative differences, inflation rate, residual value of equipment, and length of development.
- The effects of changing grade plate assumptions

Chapter 9
Step 8 in the CBA Process
Report Results and Recommendations

9.1 Documenting the CBA

A CBA preparer should document the CBA, including all tables, charts, and diagrams, according to the 8-Step Method discussed in this guide (see CBA Case Study in Appendix B for an example) preferably using a word processing application such as Microsoft Word. A CBA presented in PowerPoint is also acceptable, but it must be as thorough and comprehensive as if the CBA were prepared in Word. Ideally, the analyst should prepare a CBA using Word or similar application and then use PowerPoint to facilitate a briefing for the decision maker. A suggested format for a set of briefing slides has been included later in this section of the guide. The actual format and content of a briefing should be determined by several factors, to include the nature of the content, the briefing style of the briefer, and the preferences of the decision maker being briefed. It is beyond the scope of this guide to mandate what should or should not be briefed to a decision maker.

It is essential to thoroughly document the CBA. There must be sufficient documentation of all assumptions, costs, methodology, results, and data to enable a person unfamiliar with the project to arrive at the same conclusion as the person who prepares it. All documentation, including all supporting spreadsheets and calculations attached separately must accompany the CBA document and charts when they are submitted for review.

CBA documentation should describe the functional process performed, define the requirement, and identify significant assumptions, constraints, and key variables. The CBA documentation should also identify feasible alternatives, and present total costs and differential savings expected in constant, discounted, and current dollars over the project life. The CBA must address estimating methods/relationships and data sources, treat sensitivity, risk, and uncertainty of key cost drivers and assumptions, and address all quantifiable benefits as well as any Difficult to quantify benefits influencing the recommended course of action. Furthermore, clearly document all alternatives and the differences between them to include the justification for their deletion.

Documentation supporting the results of the analysis must include the computations, data sources, and methodologies used to estimate the costs and benefits. For example, if cost factors are used, indicate their source and/or the basic assumptions used in their derivation. All data sources should be specifically identified for all costs and benefits. Support documentation should be sufficient to allow an independent reviewer to recreate the estimate and reach the same conclusions. In addition, it is important to identify the sources of benefit data, methods used to collect the data, and quality of data. All costs for the entire project life, beginning with the first fiscal year in which costs will be incurred should be presented. Cost estimates must reflect the Army's true requirement for a system or project, not just available funding.

A recommendation as to the preferred alternative, with all appropriate supporting analysis, should accompany the comparison of alternatives.

In addition to a recommendation, an executive summary should be prepared and inserted at the front of the CBA.

Chapter 10
Summary and Conclusions

We have present a CBA analysis designed for energy security that is based on Army (U.S. Army, 2012) guidance. The CBA guidance uses a combination of well-understood engineering economic and decision analysis techniques. However, quantifying the value of energy security projects continues to be a challenge. We have presented a very simply methodology to quantifying the value using MODA techniques. Other techniques can also be used such as DEA. Using the methodology we have developed a detailed example of how to not only justify but rank order investments in energy security projects.

Chapter 11
Bibliography and References

11.1 Bibliography

Aimone, Michael A. "Cyber Warfare and the US Electric Grid Implications for National Security." Headquarters U.S. Air Force, 08 November 2010

Army Science Board, "Installations 2025 Study Report". *Version 2.4*, pp 37-43, 2006

Army Senior Energy Council "Army Energy Security and Implementation Strategy", 13 January 2009

Brockhoff, K. "On the Quantification of the Marginal Productivity of Industrial Research by Estimating a Production Function for a Single Firm", German Economic Review Vol. 7, pp. 202-229, 1970

Chouhdry, Ali, "Improving Security at the Untied States Military Academy," United States Corps of Cadets, West Point, 28 April 2010

Department of the Army AR 11-18, The Cost and Economic Analysis Program, 31 January 1995

Department of the Army AR 70-1, Army Acquisition Policy, April 2009

Department of the Army AR 700-127, Integrated Logistics Support, 26 March 2012

Department of the Army ASA(ALT) Memorandum, Performance-Based Logistics (PBL) Business Case Analysis, 23 January 2004

Department of the Army Pamphlet (DA PAM) 70-3, Army Acquisition Procedures, April 2009

DASA (Cost and Economics Directorate), Economic Analysis Manual, February 2001

DASA (Cost and Economics Directorate), Cost Management Handbook, April 2009

DASA (Cost and Economics Directorate), Budgetary and Cost Template to Support Legislative Proposals, March 2009

DASA (Cost and Economics Directorate), Cost Benefit Analysis Portal (https://cpp.army.mil)

Defense Acquisition Guidebook, Chapter 3, Affordability and Life Cycle Resource Estimates, Defense Acquisition University

DoD Cost Guidance and Tools (https://www.cape.osd.mil/costguidance/)

DoDI 5000.02 Operation of the Defense Acquisition System, December 2008

DoDI 7041.3, Economic Analysis for Decision-making, 7 November 1995

DoD Business Case Development Guide, V2, November 2003

DoD Business Case Model for the DoD Logistics Community, September 2009

DoD Financial Management Regulation, Volume 4, Chapter 19, Managerial Cost Accounting"

Farr, John V., Systems Life Cycle Costing: Economic Analysis, Estimation, and Management, CRC Press, June 2011

Field Manual 5-19, Composite Risk Management, HQDA, August 2006

Field Manual 5-0, The Operations Process, HQDA, March 2010

Government Accountability Office Cost Estimating and Assessment Guide, March 2009 (http://www.gao.gov/products/GAO-09-3SP)

Federal Emergency Management Agency, "Risk Mitigation Series – Insurance, Finance, and Regulation Primer for Terrorism Risk Management," FEMA Publication 419, accessed at http://www.fema.gov/library/viewRecord.do?id=1562, December, 2003

Goerger, Niki C., and Driscoll, Pat, "Measuring Resiliency of Metropolitan Areas: A Systems Interdependency Framework," 74th MORS Symposium, 15 June 2006

Government Accountability Office Cost Estimating and Assessment Guide, accessed at http://www.gao.gov/products/GAO-09-3SP, March 2009

Hope, Timothy, "A Value-focused Approach to Justify the Cost of Energy Security," Military Operations Research Society Workshop, 2010

National Renewable Energy Laboratory. "Targeting Net Zero Energy at Fort Carson: Assessment and Recommendations," U.S. Department of Energy, September 2010

Office of Management and Budget Circular No. A-11, Part 7, Planning, Budgeting, Acquisition, and Management of Capital Assets, accessed at http://www.whitehouse.gov/sites/default/files/omb/assets/a11_current_year/s300.pdf, August 2011

Office of Management and Budget Circular A-94, Guidelines and Discount Rates for Benefit-cost Analysis of Federal Programs, accessed at ww.whitehouse.gov/omb/circulars_a094, 29 October 1992

Security, Energy, Environmental, and Encroachment (SEEE) Panel. "Installation 2025." U.S. Army, 23 July 2009

The Secretary of Defense Memorandum, "Consideration of Costs in DoD Decision-Making" dated 27 December 2010

11.2 References

Alsfelder, George, Hartong, Timothy, Rodriguez, Michael and Farr, John V., "Methodology for Prioritization of Investments to Support the Army Energy Strategy for Installations," Center for Nation Reconstruction and Capacity Development Technical Report, Report 2012-3, DTIC: AXXXXXXX, August 2012

Department of Energy, Federal Energy Management Program, "Performing Energy Security Assessments - A How-To Guide for Federal Facility Managers," accessed at http://www1.eere.energy.gov/femp/pdfs/energy_security_guide.pdf, accessed 15 December 2011, January 2006

Hughes, Larry "Quantifying energy security: An Analytic Hierarchy Process Approach," presented at the Fifth Dubrovnik Conference on Sustainable Development of Energy, Water, and Environment Systems in

Dubrovnik, Croatia, September 2009, accessed at http://dclh.electricalandcomputerengineering.dal.ca/enen/2009/ERG200906.pdf, December 15, 2011

Keeney, Ralph L., *Value Focused Thinking*, Harvard University Press, Cambridge, Massachusetts, 1992

Kirkwood, C., *Strategic Decision Making: Multiobjective Decision Analysis with Spreadsheets*, Belmont, CA: Wadsworth Publishing Company, 1997

Nautilus Institute for Security and Sustainable Development, "Synthesis Report for the Pacific Asia Regional Energy Security (PARES) Project, Phase 1 Framework for Energy: A Framework for Energy Security Analysis and Security Analysis and Application to a Case Study of Japan," June 9, 1998, Working Draft accessed at http://oldsite.nautilus.org/archives/pares/PARES Synthesis Report.PDF, 15 December, 2011

Ramirez-Marquez, Jose Emmanuel, and Farr, John V., "Decision-making Approach for Catastrophic Scenario Selection in Disaster Recovery Planning," International Journal of Decision Support and System Technology, Volume 1, Number 2, pp 36-51, April - June, 2009

US Army, "Cost Benefit Analysis Guide," 3rd Edition, prepared by the Office of the Deputy Assistant Secretary of the Army (Cost and Economics), 22 February 2012

Appendix A
Demonstration Study Projects at Fort Bragg

Scenario[6]

We selected two hypothetical examples at Fort Bragg military installation as a means to illustrate how to develop a CBA. These projects include two combinations of installing an energy management system for the Joint Special Operations Command, an underground (URD) conversion of building E-2929, and a backup generator. An example CBA will be conducted on this project portfolio. At the end these two projects will be compared to demonstrate how to prioritize based upon economic and hard to quantify benefits from a portfolio perspective.

Project 1 – Micro Grid for the Joint Special Operations Command

Step 1: Define the Problem/Opportunity; Describe the Background

- **Problem Statement:** Fort Bragg stands as one of the largest and most densely populated military installations in the United States today; however, more importantly, it is home to U.S. Army Special Forces, 82 Airborne Division, Pope Air Force Base, U.S. Army Forces Command, U.S. Army Reserve Command, and other important tenants. The Joint Special Operational Command (JSOC) is charged to study special operations requirements and techniques to ensure interoperability and equipment standardization, plan and conduct special operations exercises and training, and develop joint special operations tactics. With the current global war on terror this is a key enabler of our military operations worldwide. In the event of a major power outage or a cyber attack this facility needs to operate.
- **Objective:** To inform agencies involved in the PEG/POM process about the needs, costs, and values of energy security projects at JSOC facilities, Fort Bragg, North Carolina, and the U.S.
- **Background:** The global war on terror has increased the emphasis on Special Forces operations. JSOC is the joint headquarters to study special operations and provide a unified command structure. To ensure that the U.S. has the ability for this command to operate uninterrupted, investments must be made in energy security to provide this capability without interuption.

Step 2: Define the Scope/Formulate Facts and Assumptions

A smart grid implementation is needed with the ultimate goal of a micro grid implementation for JSOC facilities at Fort Bragg. This is a Phase I implementation which focuses on implementing a power management system, burying the power loop that supplies JSCO facilities, putting all JSOC facilities on this loop, and/or installation a natural gas power generator. The design will be modular and easily upgradable as renewable energy and battery storage become more cost effective. However, this Phase I implementation will be the first step in developing an energy secure JSOC. By putting all JSOC facilities on the same loop and implementing smart grid technology and possibly including backup generators will dramatically increase energy security for this key organization.

Step 3: Define Alternatives

Below are three alternatives for consideration:
1) Energy system manager, place all JSOC facilities on the same power loop, and bury the power loop.
2) Energy system manager, place all JSOC facilities on the same power loop, bury the power loop, provide a backup gas power generator for the JSOC facilities.
3) Do nothing.

The second and third order effects are limited to the second alternative – the system with the generator. This generator will be tied into the local natural gas line. Natural gas is significantly more secure than

---

[6] This projects are examples that have been contrived for demonstration purposes only.

commercially generated power. However, in a large scale natural disaster or major attack the source of power for the generator could be an issue.

Step 4: Develop Cost Estimate for Each Alternative

The following table presents the cost estimate for each alternative.

Alternative	Description	Year 0 Costs	O&M (Monthly)	Salvage @ Year 20	Monthly Cost Savings	TOC
1	System without Generator	$2,000,000	$100	$0	$800	$1,832,000
2	System with Generator	$3,500,000	$500	$5,000	$800	$3,423,000
3	Do Nothing	$0	$0	$0	$0	$0
	Assume a rate of growth = interest rate thus TOC = NPV					

Year 0 costs were obtained from manufacturers data. Monthly savings were also provided by manufacturers of similar results for the system envisioned for this project. Maintenance costs are part of the contract for the energy system's manager. The $300 difference for the two options is attributed solely to diesel and maintenance of the generator.

Step 5 - Identify Quantifiable and Non-Quantifiable Benefits

The quantifiable economic benefits are presented in the table above. The difficult to quantify benefits for the various alternatives are shown in the table below.

Alternative	Metric	Value[7]	Discussion
1 – System Without Generator	Security	10	As configured this provides low impact on energy security – it is not decoupled from the grid and provides no power backup
	Compliance with Laws and Mandates	0	Not applicable
	Renewable Energy	0	Not applicable
	Energy Consumption	20	At the installation level this would reduce energy by 0.7%[8]
	Environmental Impacts	0	Not applicable
	Social/Public Perception	10	Some good publicity because of the usage of smart and micro grid technology
2 – System With A Generator	Security	40	High Impact, temporary duration, at the systems level – see Table 6.2
	Compliance with Laws and Mandates	0	Not applicable
	Renewable Energy	0	Not applicable
	Energy Consumption	20	At the installation level this would reduce energy by 0.7%
	Environmental Impacts	0	Not applicable
	Social/Public Perception	10	Some good publicity because of the usage of smart and micro grid technology

---

[7] See Section 6.3 for how these values were ascertained
[8] Where poss ble, quantifiable values should be used in justifying value.

Define Alternative Selection Criteria

The following MODA criteria were used to score alternatives.

Criteria	Weight	COA 1 (System w/o Generator)			COA 2 (System with Generator)		
		Data	Rating	Score	Data	Rating	Score
Operational Security	.60	Table 6.2	10	6	Table 6.2	40	24
Renewable Energy	.1	NA	0	0	NA	0	0
Reduced Energy	.1	Reduces Energy .7%	20	2	Reduces Energy 0.7%	20	2
Environmental Impacts	.1	Neutral	0	0	Neutral	0	0
Social/Public Perception	.1	Poor	10	1	Poor	10	1
Total Score	1.00			9.0			27.0

Step 7 - Compare Alternatives

As shown in the following table we are paying roughly $1.6M for improved economic security and the system with the generator is the best choice from a CBI perspective.

Alternative	Economic Benefits	Non Economic Benefits	CBI
1 - System w/o Generator	$1,832,000	9	203,556
2 - System with Generator	$3,423,000	27	126,778
2 – Do Nothing	$0	0	NA

The figure below shows these results graphically.

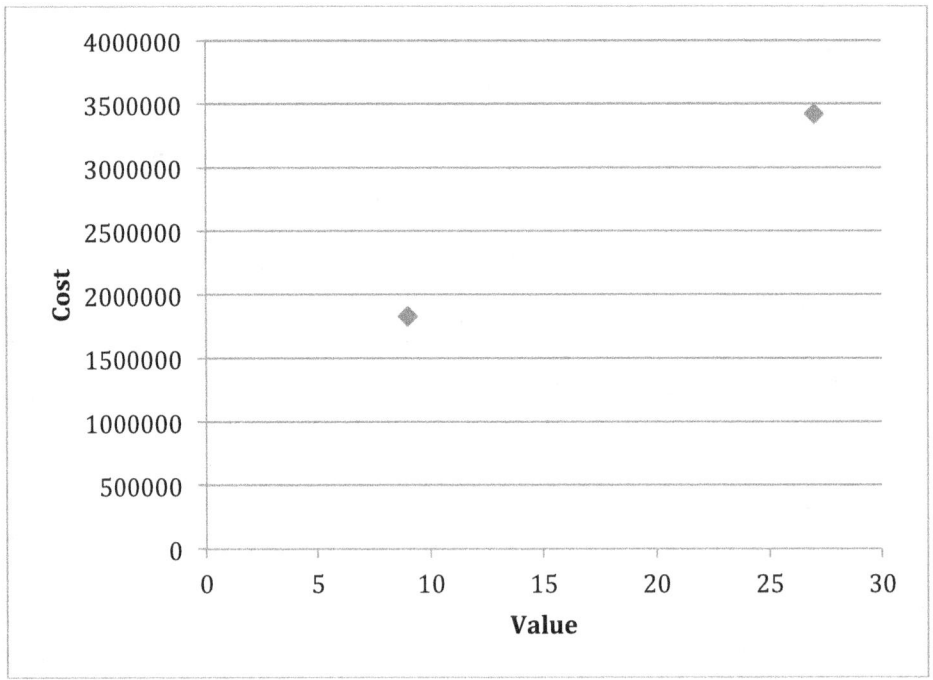

Step 8 - Report Results and Recommendations

Given the strategic importance of JSOC the additional $3.4M investment is warranted for the increase in energy security. Also, given that this is Phase I of a multi phase micro grid project this investment is needed to invest in the total program. We believe that within the next 5 years as renewable energy and storage becomes more cost effective and reliable that JSOC will be operational with a micro grid that is totally self sufficient such that in a time of major crisis (i.e., the Fort Bragg grid is not operating) that is can still carry out its basic functions. This phase I investment is needed as a bridge to phase II and to provide some means of near term energy security in the event of hurricanes, power outages, etc.

www.ingramcontent.com/pod-product-compliance
Lightning Source LLC
Chambersburg PA
CBHW081410170526
45166CB00010B/3279